Hour Telephone Renewals 020 8489 4

HARINGEY LIBRARIES

ST BE RETURNED ON

NJ

D0419242

The Genius Of THE ROMANS

CLEVER IDEAS AND INVENTIONS FROM PAST CIVILISATIONS

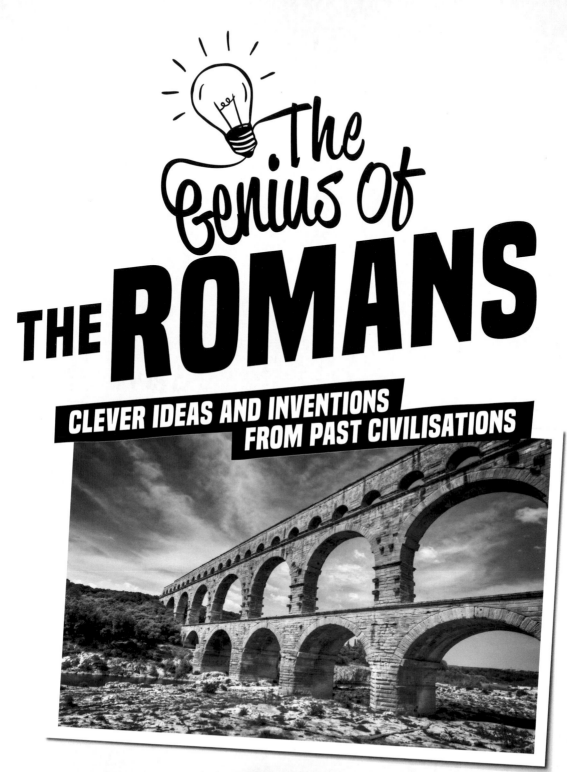

IZZI HOWELL

W

FRANKLIN WATTS
LONDON • SYDNEY

Franklin Watts

First published in Great Britain in 2019 by
The Watts Publishing Group

Copyright © The Watts Publishing Group, 2019

Produced for Watts by
White-Thomson Publishing Ltd
www.wtpub.co.uk

All rights reserved.

Series Editor: Izzi Howell
Consultant: Philip Parker
Series Designer: Rocket Design (East Anglia) Ltd

ISBN: 978 1 4451 6112 9 (HB)   978 1 4451 6113 6 (PB)
10 9 8 7 6 5 4 3 2 1

FSC
MIX
Paper from
responsible sources
FSC® C104740
www.fsc.org

Franklin Watts
An imprint of
Hachette Children's Group
Part of The Watts Publishing Group
Carmelite House
50 Victoria Embankment
London EC4Y 0DZ

An Hachette UK Company
www.hachette.co.uk

www.franklinwatts.co.uk

Printed in China

Picture acknowledgements:
Alamy: Chronicle 15b, North Wind Picture Archives 17, Atlaspix 22, nicoolay 27, The Print Collector 29;
Getty: 1001nights cover, Neil Holmes 6, DEA / G. DAGLI ORTI/De Agostini 9t, mmac72 10, georgeclerk 11,
ANDREAS SOLARO/AFP 14, Aksenovko 15t, AndreaAstes 16, DEA / L. PEDICINI 19b, Stefano Bianchetti/Corbis 25;
Shutterstock: Lagui 3, McCarthy's PhotoWorks 4l, rusty426 4r, Sopotnicki 5b, Marco Ossino 9b, Brian Maudsley
12, bigacis 18t, MIGUEL GARCIA SAAVEDRA 18c, Volosina 18bl, baibaz 18br, eFesenko 19t, Mirec 20, Giakita 21,
Kamira 23, S.Borisov 24, Algol 26t, J. Lekavicius 26b, meunierd 28, Marco Rubino 30.

All design elements from Shutterstock.

Every effort has been made to clear copyright. Should there be any inadvertent omission,
please apply to the publisher for rectification.

The website addresses (URLs) included in this book were valid at the time of going to press.
However, it is possible that contents or addresses may have changed since the publication of this book.
No responsibility for any such changes can be accepted by either the author or the publisher.

HARINGEY PUBLIC LIBRARY

70003008237 8	
PETERS	12-Feb-2019
£12.99	

CONTENTS

THE ROMANS

Who?

The Romans were one of the greatest and most influential civilisations of all time. They started in around 1000 BCE as a group of Latin-speaking people who lived on the banks of the River Tiber (which flows through the city of Rome in Italy). By 600 BCE, this group had become an important city-state, ruled by kings.

The Romans then went through different stages of rule. First, there was the Roman Republic, followed by the Roman Empire and the rule of emperors. The Romans conquered land across Europe, the Middle East and North Africa, spreading their culture and civilisation and leaving a legacy that lives on today.

The Romans carried banners like this into battle. 'SPQR' stands for 'the Roman Senate and people' in Latin.

Emperor Tiberius Caesar

According to legend, the twins Romulus and Remus were the founders of the city of Rome. The myth states that they were raised by a wolf after their parents abandoned them.

10th legion

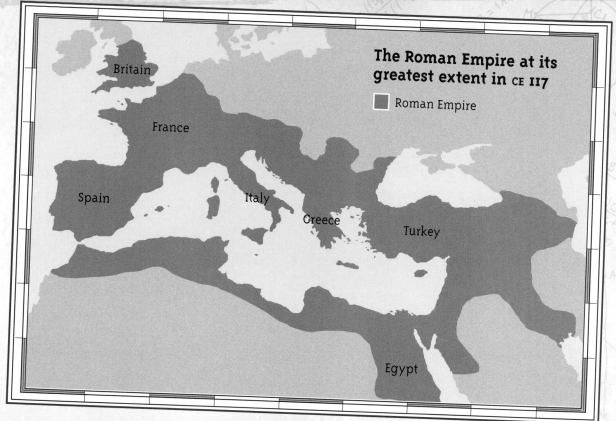

The Roman Empire at its greatest extent in CE 117

◼ Roman Empire

Britain

France

Spain

Italy

Greece

Turkey

Egypt

Millions of tourists visit the Colosseum in Rome every year, which the Romans built for gladiator matches (see pages 28–29).

What happened?

Over time, the Roman Empire was weakened by civil war and invasions from neighbouring tribes. Eventually, it split in half. The Western Roman Empire ended and other groups seized control of its territory. The Eastern Roman Empire, which became known as the Byzantine Empire, continued for over 1,000 years in the Balkans and the Middle East. Today, Roman influences and remains can still be seen across Europe, North Africa and the Middle East.

THE ARMY

The Romans' large and organised army was key to the expansion of its empire. Around the 1st centuries BCE and CE, the Roman army went from strong to unbeatable. This was thanks to to one change: being a soldier became a full-time job.

★ GENIUS ★ WELL-TRAINED PROFESSIONAL SOLDIERS

Volunteer to professional

Ancient armies were made up of volunteer soldiers, who only fought when needed. They provided their own weapons and uniform. However, professional soldiers in the Roman army were committed to military life. They signed up for around twenty years of service and received a salary. The army provided standard armour and weapons, such as the short *gladius* sword, so that everyone was equally prepared for battle.

Training

Having a professional army meant that the Romans could invest in training, as they knew that the soldiers would be around to fight in battle. New army recruits had to train hard, working on their fitness, learning to follow commands and work as a team, and practising fighting with swords and javelins. A Roman soldier was expected to walk 30 km a day in full armour and build a new camp every evening.

These re-enactors show how Roman soldiers held their shields to protect themselves from enemy spears and arrows.

Into battle

The soldiers' training and preparation paid off on the battlefield. Rather than running straight into battle, foot soldiers, archers and cavalry (soldiers on horseback) moved tactically, working together to take down the enemy. Thanks to their hours of practice, they knew how to quickly get into different military formations on the command of their leaders.

WOW!

Any bad behaviour could affect the organisation and teamwork of the army, so discipline was important. A soldier who misbehaved was punished by fines, dismissal or even death.

BATTLE PLANS

The army's tactics depended on the size and strength of the enemy and the geography of the battlefield.

enemy troops

Roman soldiers

1 Wedge Formation
Soldiers concentrated in the centre to break through enemy lines

enemy troops

Roman soldiers

2 Single line defence
Soldiers arranged in a single line that wraps around enemy troops

enemy troops

Roman soldiers

3 Weak centre
Few soldiers placed in the centre to tempt enemy to attack there, after which remaining soldiers drive in on either side

))) BRAIN WAVE)))

In the *testudo* (tortoise) formation, soldiers packed tightly together to form a square, using their shields like a tortoise shell to protect their sides and heads. The result was an impenetrable block of soldiers, who could move together towards the enemy without getting hurt.

7

TRADE

The Romans controlled many different areas (see page 5). They conquered these lands to expand their empire and power. Their huge gains in territory came with the added bonus of access to different resources.

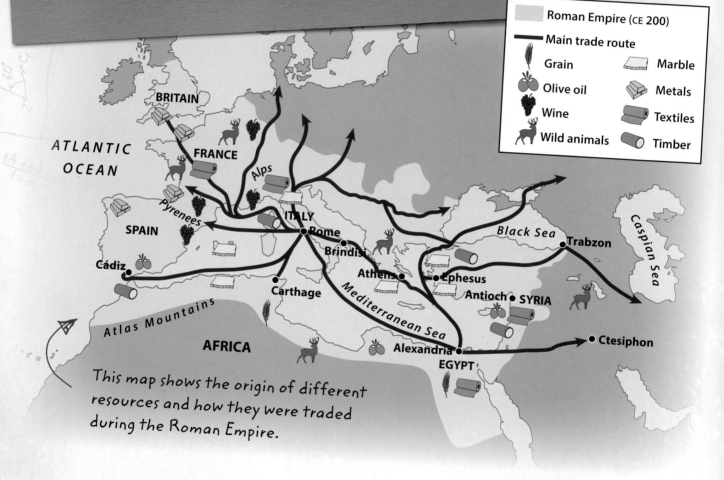

	Roman Empire (CE 200)
	Main trade route

	Grain		Marble
	Olive oil		Metals
	Wine		Textiles
	Wild animals		Timber

This map shows the origin of different resources and how they were traded during the Roman Empire.

Across the empire

Roman territory included lands with varying climates and natural habitats. Some resources, such as lead, could only be found in certain areas, such as Britain and Spain. The Romans transported this lead to Italy, where it was used to construct water pipes in Rome. Other regions had climates that were well suited to growing certain crops. Farmers in North Africa grew wheat that was taken across the empire and used to make bread.

Road or sea

The wide-reaching Roman road network (see pages 12–13) made it easier for traders to move items around the empire. However, the Romans worked out that it could sometimes be cheaper and quicker to send products across the Mediterranean. They built large ports where products could be received and sent inland. The port of Ostia was built on the coast near to Rome to supply the city with traded goods.

Luxury goods

Roman traders established trade connections beyond their empire, giving them access to an even wider range of products. Traders sailed across the Indian Ocean to India, bringing back luxury items such as silk, perfume, pepper and cinnamon. The time and effort required to collect these items would have made them extremely expensive.

(((BRAIN WAVE)))

Pirates often disrupted trade in the Mediterranean Sea. This could affect food supply across the empire. In 67 BCE, the politician Pompey was given control over the Mediterranean Sea and coastline, and a large fleet of ships to fight the pirates. He defeated them in just a few months.

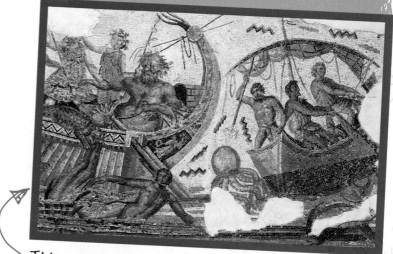

This mosaic shows the Roman god Dionysus fighting off pirates.

The Romans also traded wild animals with India. This Roman mosaic shows tigers, which can only be found in Asia. Tigers were brought to Italy for gladiator shows (see page 29).

CONCRETE

The Romans developed a new recipe for concrete, which was strong and versatile. They used this concrete to construct arched aqueducts and bridges, and massive domed roofs.

★ GENIUS ★
NEW RECIPE

A Roman recipe

Roman concrete was made from small chunks of rock held together by mortar made with volcanic ash and seawater. This mixture set into a hard block that wasn't just resistant to damage, but actually got stronger over time when placed in the sea. Modern scientists have discovered that the reaction between volcanic ash and seawater created new minerals that filled in any cracks in the concrete. Some types of Roman concrete could even set underwater, making it an ideal material for harbours, bridge supports and sewers.

The Romans used volcanic ash mortar to bind together the top blocks of the Pont du Gard aqueduct in the south of France. The rest of the blocks were cut so accurately that they didn't need mortar to stay in place.

New shapes

Roman concrete could be poured into moulds to create building blocks in any shape. Previously, builders had to chisel pieces of stone down to size, which was time-consuming and required high levels of skill. The Romans didn't like the appearance of concrete, so they covered concrete blocks in bricks or stone.

Daring domes

The greatest Roman concrete dome is the roof of the Pantheon temple in Rome. It is 43 m in diameter and 22 m in height, supported only by the wall beneath it. Previously, roofs had been supported by internal pillars, which took up a lot of space inside the building. But not here — the inside of the Pantheon is a vast, open space, with the domed roof rising unsupported above.

TEST OF TIME

As modern concrete can break down in as little as 50 years, some architects are considering using Roman concrete for new building projects, such as a proposed tidal power plant in Swansea, Wales.

This oculus (hole) in the ceiling of the Pantheon lets natural light stream into the building.

WOW!

To this day, the Pantheon is the largest unreinforced concrete dome in the world. It does not contain any metal supports — only concrete.

ROADS

While the Roman army helped to expand the empire, the construction of roads helped to maintain it. The Romans built a vast road network linking different parts of their territory, which made communication and trade much easier.

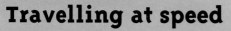

★GENIUS★
NETWORK OF STURDY STREETS

Travelling at speed

Having a large network of well-built roads allowed soldiers and supplies to move quickly around the empire. Messengers with the official Roman postal service could travel as much as 270 km in a day and a night. If there was a rebellion against the Romans, the army could easily reach the rebels by travelling along Roman roads. This helped to keep conquered lands under control and maintain the empire.

This Roman road goes through the mountains of Spain. The Romans preferred to build roads on high ground, as they were less likely to be flooded. People travelling along the road were also safer, as they could see anyone approaching.

WOW!

At their peak, the Romans had built over 85,000 km of main roads, connecting the city of Rome with different parts of the empire.

Planning a road

The Romans were particular about the location of their roads. After they conquered new territory, they would send in map makers to study the area and plan routes for new roads. They built roads to link strategic points, such as towns and cities. For example, the Via Appia linked Rome to Brindisi, an important port. Most Roman roads were built in straight lines, as this covered the least distance.

TEST OF TIME

Many Roman roads still exist today. In some places, the original surface can still be seen. In other areas, new roads have been built on top of Roman roads, following the same direct route.

Construction

The construction of a Roman road depended on its use. Busy roads were built to be much stronger than less-used roads. Road builders began by digging two shallow ditches on either side of the path of the road. Then, they built up the road above ground level between the ditches. The road was made from layers of compacted sand, crushed rock and concrete, topped with large flat stones.

large stone slabs

cemented sand and gravel

gravel in cement mortar

compacted sand or dry earth

The road would slope down from the middle to the sides, so that water would run off into the ditches at the sides.

WATER

The Romans needed large quantities of water to supply cities and public baths. Luckily, they were excellent engineers, building aqueducts to bring clean water into their cities, and sewers to take dirty water away.

WOW!

It has been estimated that Rome's eleven aqueducts brought millions of litres of clean water to the city every day!

The Acqua Vergine aqueduct still brings water into Rome. It provides the water for the Trevi Fountain.

Bringing water

Aqueducts meant that the Romans didn't have to depend on a good local fresh water supply. They could build cities in dry areas without much water. They could also supply large cities, such as Rome, with enough fresh water to support the population. In Rome, fresh water from aqueducts went into public fountains and drinking taps, where ordinary people collected water for everyday use. The wealthiest Romans had their own private water supply connected to their house.

))) BRAIN WAVE)))

The Romans built zigzags into some aqueducts. This slowed down the water so that it was more likely to drop its muddy sediment. This meant that the water was cleaner when it reached its destination.

Waste and sewage

The Romans built one of the earliest sewer systems in Rome, called the Cloaca Maxima. At first, the Cloaca Maxima was an open channel that drained water from the marshland where the Forum was built (see page 24). Later, it was covered over and used to carry dirty water from baths and public toilets into the River Tiber. The Romans built sewer systems in other parts of the empire, such as Eboracum (York) in the UK.

Waste from public toilets fell into a trench underneath the seats.

Public baths

Having a reliable supply of water was vital for Roman public baths. Bathing was an important part of ancient Roman culture, and people would regularly visit public baths to wash, relax and socialise. The Romans built baths across the empire. One of the largest was the Baths of Diocletian, which was built in Rome in around CE 300. It could fit 3,000 people and had facilities such as swimming pools, warm and cold baths, steam rooms and places to exercise.

Men and women used separate sections of the baths, as shown in this modern painting.

CALENDARS

The early Romans looked to the skies to help keep track of the days, months and years. However, their first calendar, which was based on the cycles of the Moon, didn't really work. After some complicated calculations, the Romans produced a new, more accurate calendar, which is almost identical to the one that we use today.

★ GENIUS ★
LEAP YEAR

Calendar problems

The early Roman calendar based on the Moon only had 355 days in a year. This meant that it had to be constantly adjusted by adding extra days. These days were not always added and so the calendar slowly got out of sync. Important moments in the farming year no longer matched the correct calendar month. By the first century BCE, the calendar was three months out of sync.

All change

In the 40s BCE, the Roman general Julius Caesar (left) ordered that the calendar be revised. He consulted a Greek astronomer, who suggested a version of the Egyptian solar calendar. The new Roman calendar had 365 days divided across twelve months. Every four years, there would be a leap year with one extra day. Having leap years kept the calendar almost perfectly in sync.

The early Roman calendar was so out of sync that it showed the harvest to be happening in summer, rather than autumn!

Catching up

Before the Romans could start using their new calendar, they had to get the calendar back into sync. To do this, 46 BCE had to have 445 days! 45 BCE was the first 365-day year. The new calendar made it much easier to keep track of important religious festivals and dates.

The ancient Greeks and Egyptians were advanced astronomers, and the Romans benefited from their knowledge.

TEST OF TIME

The Roman calendar was used until the Gregorian calendar was introduced by Pope Gregory XIII in 1582. The Gregorian calendar adjusted the Roman calendar by a fraction, making it even more accurate. Some Orthodox Christian Churches still use the Roman calendar to calculate the date of Easter.

WOW!

The fifth month in the new calendar was named 'Julius' after Julius Caesar. Today, we call this month July.

FOOD

There were lots of hungry mouths to feed in Rome and across the Roman Empire. Roman leaders realised that making sure people were fed helped to keep the peace. However, the food that people ate depended greatly on their income.

Dates

An edible empire

The Romans ate foods from every region of their empire. Rich Romans enjoyed dates and pomegranates from northern Africa, high-quality olive oil from Spain and spices from Asia. The Romans also collected grain from across the empire to feed the poor. Many of these foods had to be preserved so that they wouldn't go off while being transported. The Romans pickled fruit and vegetables and salted meat so that they would last the journey.

Olives

Wheat

WOW!

Romans used *garum*, a sauce made from fermented fish, as we use ketchup. Historians think that *garum* factories were built outside cities, as the smell of the fermenting fish was so disgusting that people couldn't live nearby!

Pomegranate

Everyday food

Most ordinary Romans could not afford an extravagant and varied diet. They mostly ate barley, wheat, vegetables and local olive oil, with occasional pieces of meat. This was a fairly balanced diet that kept people healthy. In cities, most ordinary people lived in *insulae* — flats with no kitchen that just had room for a one-pot stove over a fire. They could only cook simple meals at home.

(((BRAIN WAVE)))

In cities, innkeepers and shopkeepers saw an opportunity to sell fast food to ordinary people who didn't have space to cook at home. They sold cheap, ready-to-eat meals, such as stews and sausages. Romans could eat in or take away to eat at home.

Political food

Sometimes, food could be hard to come by, so politicians handed out food to win votes and approval. They tried to persuade poor people to vote for them by lowering the price of grain and even handing out free grain. When they stopped giving out food, poor people protested. The politicians quickly starting giving out food again. There were many poor people in Rome, so it would have been risky to make them all angry and rebellious!

This *thermopolium* (fast-food restaurant) in Pompeii is so well preserved that you can still see the holes on the counter that were filled with jars containing food.

This Roman fresco shows bread being distributed. During the reign of emperor Aurelian (CE 270–275), bread was given out instead of grain.

19

THE LATIN LANGUAGE

During the Roman age, Latin grew into the language of an entire empire. After the end of the Roman Empire, Latin continued to be used in science, as well as developing into new languages.

Spoken and classical

The Romans used two different versions of Latin — spoken and classical. Each version had different words and grammar. Spoken Latin was used by ordinary people for everyday conversations and was rarely written down. Modern Romance languages, such as Italian, developed from spoken Latin.

Classical Latin was taught at school and used to write important documents. The Romans created many different written works, ranging from stories to philosophical essays, historical accounts and poetry. Much of what we know about ancient Rome is taken from these texts.

This engraving at the Colosseum (see page 5) in Rome is in classical Latin. We know much more about classical Latin than spoken Latin, as we have more written examples to study, such as the text on monuments.

Coming together

People in the Roman Empire were not required to speak Latin to become a Roman citizen. However, many people who lived in lands that came under Roman rule chose to learn Latin. Latin was used for government administration and laws, so knowing it helped new people in the empire to integrate into Roman society. A shared language allowed people from across the Roman Empire to communicate with each other and feel united as one group of people.

WOW!

The Latin alphabet is still used to write English and many other languages around the world. Today, it is the most widely used writing script in the world.

TEST OF TIME

Today, many scientific, legal and medical terms have Latin roots or are still written in Latin. For example, doctors use the phrase nil by mouth to describe a patient who shouldn't eat anything. 'Nil' means 'nothing' in Latin.

A replica of a Roman wax tablet. Archaeologists have found a few wax tablets with messages in spoken Latin, which have given us some clues about the language.

Writing it down

The Romans didn't have paper. They wrote on parchment made from papyrus plants. Papyrus parchment was very expensive to produce, so it was only used for the most important documents. The Romans came up with a clever solution for everyday writing. They filled wooden frames with solid wax and used a tool to carve messages into them. Later, they could melt the wax and erase the message, leaving the tablet ready to be used again.

GOVERNMENT

During the 500-year-long Roman Republic, the Romans were ruled over by a government made up of two consuls and a council called the Senate. This system stopped one ruler from becoming too powerful.

GENIUS ★ POWER-SHARING REPUBLIC

Changing consuls

The two elected consuls were in charge of the government and the army. To stop the consuls becoming too powerful, they only ruled for one year each. After 367 BCE, the Romans made things even fairer by deciding that at least one of the consuls had to be from an ordinary family rather than a wealthy one.

The Roman consul Cicero talks to members of the Senate. Cicero served as consul in the year 63 BCE.

TEST of TIME

The Roman Republic inspired the French Revolution (1787–1789), in which the French people overthrew the monarchy and started their own republic. France remains a republic today, ruled over by a president and a government. It does not have a monarchy.

The Senate

The consuls received advice from the Senate, a council of around 300 men. The Senate members came from the most important families in Rome. This gave them a huge amount of power, as they had many well-connected relatives and friends, who the consuls did not want to annoy! For this reason, the consuls almost always took the advice of the Senate.

Republic to Empire

In 27 BCE, the Roman Republic became the Roman Empire. It was ruled over by a series of emperors. The emperor held all the power and made his own decisions about the empire. The Senate still existed, but it didn't have much power. Unlike during the Republic, the Senate was controlled by the emperor, rather than the Senate controlling the consuls. Some Roman emperors managed the empire well, while others were corrupt and used their position for their own personal benefit.

WOW!

The emperor Caligula (ruled CE 37 to 41) is famous for his cruel and strange behaviour. He is reported to have ordered soldiers to attack the sea and collect seashells to show that he had defeated the ocean!

Antoninus Pius was the most peaceful Roman emperor. There were no major wars during his rule (CE 138–161). He managed money well and built new temples.

(((BRAIN WAVE)))

Although the emperor was the official head of the Roman Empire, governors ruled over sections of the empires, known as provinces. Each governor controlled the army in his province and acted as a judge. This freed up time for the emperor to focus on the most important issues.

LAWS

Roman society was not very equal. Women, slaves and people who weren't Roman citizens did not enjoy the same treatment as rich men. However, the Romans did create a system of laws that made things slightly fairer and helped to stop the wealthy from abusing their power.

★ GENIUS OFFICIAL LAWS ★

Unfair laws

In the early Roman Republic, only people from rich families could be magistrates. They made up the laws and chose how and when to use them, deciding if someone was guilty and what punishment they should receive. This gave them a lot of power over ordinary people. They often abused the law to benefit themselves, make money and punish their enemies.

The Twelve Tables were displayed in the Forum in Rome. During Roman times, this was a large public square between important government buildings.

The Twelve Tables

In 451 BCE, the Romans created the first law code, called the Twelve Tables. This was a list of twelve laws, written on bronze tablets and displayed publicly in Rome. These laws applied to everyone. As they were written down and could be seen by anyone, it was harder for rich people to change them. Over time, the Romans added more laws to their law code, covering many aspects of life and business. The rules on who could be a magistrate also changed to include any free man, rich or poor.

Punishments

Romans gave out serious punishments to people accused of breaking the law. They hoped that these punishments would be a warning to others who were tempted to break the law. Many crimes were punished by death. However, the way in which people were killed depended on who they were. Wealthy people sentenced to death were executed in private, while ordinary people were killed publicly. Slaves were killed during gladiator matches (see page 29) or by crucifixion, the same way in which Jesus was killed.

(see page 29)

(((BRAIN WAVE)))

The new law code created the need for a new job that had never existed before – the lawyer. Roman lawyers helped people to interpret and understand Roman law. Some Romans also worked as law experts, who studied laws and tested them against imaginary cases to make sure they were working well.

One Roman form of execution was to throw the person sentenced to death off a cliff.

CITY SERVICES

In a large city, such as Rome, the risk of fire and crime was fairly high. The world's first fire department helped to stop fires and keep criminals off the streets.

The lower floors of insulae were more expensive as they had larger rooms and were less of a fire risk.

At home

Many people in Rome lived in small flats (*insulae*) in buildings in narrow, crowded streets. The bottom floors of these apartment blocks were made from brick or stone, while the top floors were made from wood. People in these flats used oil lamps for light, and open fires for cooking and heat. For people living in the top wooden floors, this was a serious fire risk. What's more, the buildings were so close together that a fire in one flat could have quickly spread into a huge citywide fire.

Romans burned plant oil, such as olive oil, in their oil lamps.

Vigiles

To prevent the risk of a serious fire in Rome, the emperor Augustus started a fire department in CE 6. The fire department employed professional firemen called *vigiles*. Their name came from the Latin word 'vigil', meaning alert and watchful. Over 7,000 *vigiles* patrolled the streets of Rome, looking out for fire.

On patrol

While the *vigiles* weren't putting out fires, they had time to patrol the city at night, acting as policemen. They kept an eye out for thieves and runaway slaves, and sorted out low-level crimes such as fighting. This made the streets of Rome safer.

(((BRAIN WAVE)))

In case of a fire, *vigiles* could use a 'fire engine' to put out the flames. This was a large water pump on a wagon, pulled by horses. They used the pump to get water from a water source near the fire. Then, they brought buckets of water to put out the fire.

TEST OF TIME

The *vigiles* were the world's first fire department. Today, over two thousand years later, almost every town and city has a fire department with firefighters and state-of-the-art equipment to put out dangerous fires.

The emperor was protected by soldiers from the Praetorian Guard. In the case of a serious crime, the Praetorian Guard could be called on to help the vigiles.

SHOW TIME!

Roman emperors were keen to keep poor people entertained, as they were less likely to revolt when they were distracted. They built huge buildings where free entertainment was put on, such as chariot races and gladiator fights.

GENIUS ★ ENTERTAINERS ★

Big buildings

The buildings where entertainment was put on had to be large enough to fit huge crowds of spectators. In Rome, the Colosseum (see page 5) could seat over 50,000 people, while the Circus Maximus stadium had space for over 150,000 people. The Romans also built large amphitheatres, racetracks and theatres in other parts of the empire.

Risk of danger

Roman entertainment was exciting and dangerous to keep people interested. There was a high risk of accident in many events. In Roman chariot races, twelve chariots raced seven laps around a huge oval track. Each chariot racer tried to get the best position near to the barrier in the centre. Crashes at high speed were very common.

The chariots were as light as possible to make it easier for the horses to pull them. However, this meant that there wasn't much protection for the driver if they crashed into the central barrier or another chariot.

Cory sports

Some Roman sports were even more bloodthirsty. In gladiator matches, gladiators fought to the death or until one of them surrendered. Spectators could also watch criminals receive a terrible death sentence — fighting without weapons or armour against wild animals, such as lions or bears. If the criminal managed to kill the animal, another animal was released immediately.

(((**BRAIN WAVE**)))

The Romans used advertising to attract large numbers of spectators to gladiator events. To drum up a big crowd, they painted adverts on buildings a few days before the show. The adverts mentioned the names of the gladiators who would be fighting, the style of fighting and when and where the event would take place.

WOW!

Each type of gladiator had a different fighting style, to appeal to a wide range of people. A *retarius* gladiator fought with a net and a trident, while the *andabata* gladiator is thought to have fought on horseback in a helmet with a closed visor.

A *retarius* gladiator attacks a *secutor* gladiator. The *secutor* was trained to fight the *retarius*, and carried a large shield and a short sword.

trident

29

GLOSSARY

amphitheatre — an outdoor venue with a round stage surrounded by rows of seats

aqueduct — a structure for carrying water across land

astronomer — someone who studies planets and the stars

conquer — to take control of a foreign land, usually using force

consul — during the Roman Republic, two consuls were in charge of the government and the army

dictator — a leader who has complete control of a country and has not been elected

emperor — the ruler of an empire

magistrate — a type of judge

mortar — a substance used to hold bricks or stones together

mosaic — a picture or pattern made out of small coloured tiles

parchment — an ancient type of paper made from dried animal skin

Praetorian Guard — a section of the Roman army whose soldiers were bodyguards for the emperor

republic — a country whose leader is elected by the people

revolt — when a large number of people protest against their leaders and refuse to be ruled by them

Romance language — a language that developed from Latin, such as Italian

salary — the money paid to someone for doing their job. The word salary comes from Latin and means 'the money given to a soldier to buy salt'

territory — an area of land that belongs to someone

trident — a weapon that consists of a pole with three metal points on the end

visor — the part of a helmet that can be pulled down to cover the face

The city of Rome today.

TIMELINE

753 BCE   The city of Rome is founded.

600 BCE   Rome is a city state ruled by kings.

509 BCE   The last Roman king is overthrown and the Roman Republic starts.

46 BCE   Julius Caesar takes control as a dictator.

44 BCE   Julius Caesar is murdered.

27 BCE   The Roman Empire begins and the first emperor, Augustus, starts to rule.

43 CE   Britain is invaded by the Romans.

CE 117   The Roman Empire is at its greatest size under the emperor Trajan, ruling over areas including what is now Spain, the UK, northern Africa and Turkey.

CE 235–284   Many Roman emperors are murdered, as different people try to seize power from each other.

CE 376   The Goths (Germanic invaders) invade much of the Roman Empire.

CE 395   The Roman Empire is split in two.

CE 410   The last Romans leave Britain.

CE 476   The Western Roman Empire ends.

CE 1453   The Eastern Roman Empire, later known as the Byzantine Empire, ends.

INDEX

FURTHER INFORMATION

Websites

www.dkfindout.com/uk/history/ancient-rome/
www.bbc.co.uk/education/topics/zwmpfg8
www.primaryhomeworkhelp.co.uk/Romans.html
www.pompeiisites.org/

Books

How They Made Things Work: Romans by Richard Platt (Franklin Watts, 2018)

Ancient Rome (Facts and Artefacts) by Tim Cooke (Franklin Watts, 2018)

Solving the Mysteries of Ancient Rome by Trudy Hanbury-Murphy (Franklin Watts, 2014)

The Genius Of

Titles in the series

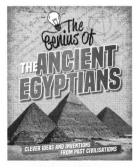

- Who were the Egyptians?
- Pyramids • Temples
- Writing • Papyrus
- Farming methods
- Irrigation • Calendars
- Clocks • Mummification
- Medicine • Toothpaste
- Cosmetics

HB 9781445161198
PB 9781445161204

- The Greeks • The Empire
- Democracy • Sports
- Medicine • Philosophy
- Warfare • Buildings
- Theatre • Science • Maths
- Art • Astronomy

HB 9781445161211
PB 9781445161228

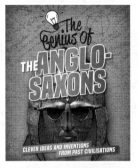

- The Anglo-Saxons
- Kingdoms and rulers
- Society • Towns • Laws
- Old English • Trade
- Art • Food • Defence
- Weapons and armour
- Entertainment • Clothes

HB 9781445161174
PB 9781445161181

- What was the Benin Kingdom? • Powerful leaders • The city-state
- Professional soldiers
- Farming • Trade • Town planning • Craft guilds
- Art • Metalwork
- Working with wood
- Textiles • Botany

HB 9781445161259
PB 9781445161266

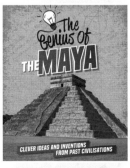

- The Maya • Government and kings • Trade • Warfare
- Cities • Buildings
- Writing • Food
- The Mayan calendar
- Astronomy • Sports
- Art • Clothes

HB 9781445161235
PB 9781445161242

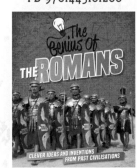

- The Romans • The army
- Trade • Concrete • Roads
- Water • Calendars • Food
- The Latin language
- Government • Laws
- City services • Show time!

HB 9781445161129
PB 9781445161136

- The Stone, Bronze and Iron Ages • Stone
- Bronze • Iron • Farming
- Construction
- Settlements • Society
- Trade • Clothing • Art
- The wheel • Writing

HB 9781445160467
PB 9781445160474

- The Vikings • The Viking longship • Sails and keels
- Compasses • Exploration
- Trade • Battle-axes
- Shields • Law and democracy • Language
- Skiing • Personal grooming • Viking sagas

HB 9781445161167
PB 9781445161143